Bug Identification Log Book

Contact Information	
Name:	
Phone:	
Email:	

BUG IDENTIFICATION

SEASON

- Spring ☐
- Summer ☐
- Fall ☐
- Winter ☐

WEATHER

- Sunny ☐
- Rainy ☐
- Foggy ☐
- Cloudy ☐

Bug Name:	
Scientific Name:	
Location:	
Colors:	
Number of Legs:	
Does It Have Wings?	
Behavior:	
Alone or in Group?	

PHOTO/DRAWING

NOTES

BUG IDENTIFICATION

SEASON

- Spring ☐
- Summer ☐
- Fall ☐
- Winter ☐

WEATHER

- Sunny ☐
- Rainy ☐
- Foggy ☐
- Cloudy ☐

Bug Name:	
Scientific Name:	
Location:	
Colors:	
Number of Legs:	
Does It Have Wings?	
Behavior:	
Alone or in Group?	

PHOTO/DRAWING

NOTES

BUG IDENTIFICATION

SEASON

- Spring ☐
- Summer ☐
- Fall ☐
- Winter ☐

WEATHER

- Sunny ☐
- Rainy ☐
- Foggy ☐
- Cloudy ☐

Bug Name:	
Scientific Name:	
Location:	
Colors:	
Number of Legs:	
Does It Have Wings?	
Behavior:	
Alone or in Group?	

Photo/Drawing

Notes

BUG IDENTIFICATION

SEASON	
Spring	☐
Summer	☐
Fall	☐
Winter	☐

WEATHER	
Sunny	☐
Rainy	☐
Foggy	☐
Cloudy	☐

Bug Name:	
Scientific Name:	
Location:	
Colors:	
Number of Legs:	
Does It Have Wings?	
Behavior:	
Alone or in Group?	

PHOTO/DRAWING

NOTES

Bug Identification

Season

- Spring ☐
- Summer ☐
- Fall ☐
- Winter ☐

Weather

- Sunny ☐
- Rainy ☐
- Foggy ☐
- Cloudy ☐

Bug Name:	
Scientific Name:	
Location:	
Colors:	
Number of Legs:	
Does It Have Wings?	
Behavior:	
Alone or in Group?	

PHOTO/DRAWING

NOTES

Bug Identification

Season
- [ ] Spring
- [ ] Summer
- [ ] Fall
- [ ] Winter

Weather
- [ ] Sunny
- [ ] Rainy
- [ ] Foggy
- [ ] Cloudy

Bug Name:	
Scientific Name:	
Location:	
Colors:	
Number of Legs:	
Does It Have Wings?	
Behavior:	
Alone or in Group?	

Photo/Drawing

Notes

Bug Identification

Season	
Spring	☐
Summer	☐
Fall	☐
Winter	☐

Weather	
Sunny	☐
Rainy	☐
Foggy	☐
Cloudy	☐

Bug Name:	
Scientific Name:	
Location:	
Colors:	
Number of Legs:	
Does It Have Wings?	
Behavior:	
Alone or in Group?	

Photo/Drawing

Notes

Bug Identification

Season
- Spring ☐
- Summer ☐
- Fall ☐
- Winter ☐

Weather
- Sunny ☐
- Rainy ☐
- Foggy ☐
- Cloudy ☐

Bug Name:	
Scientific Name:	
Location:	
Colors:	
Number of Legs:	
Does It Have Wings?	
Behavior:	
Alone or in Group?	

PHOTO/DRAWING

NOTES

BUG IDENTIFICATION

SEASON

- Spring ☐
- Summer ☐
- Fall ☐
- Winter ☐

WEATHER

- Sunny ☐
- Rainy ☐
- Foggy ☐
- Cloudy ☐

Bug Name:	
Scientific Name:	
Location:	
Colors:	
Number of Legs:	
Does It Have Wings?	
Behavior:	
Alone or in Group?	

PHOTO/DRAWING

NOTES

BUG IDENTIFICATION

SEASON	
Spring	☐
Summer	☐
Fall	☐
Winter	☐

WEATHER	
Sunny	☐
Rainy	☐
Foggy	☐
Cloudy	☐

Bug Name:	
Scientific Name:	
Location:	
Colors:	
Number of Legs:	
Does It Have Wings?	
Behavior:	
Alone or in Group?	

PHOTO/DRAWING

NOTES

BUG IDENTIFICATION

SEASON	
Spring	☐
Summer	☐
Fall	☐
Winter	☐

WEATHER	
Sunny	☐
Rainy	☐
Foggy	☐
Cloudy	☐

Bug Name:	
Scientific Name:	
Location:	
Colors:	
Number of Legs:	
Does It Have Wings?	
Behavior:	
Alone or in Group?	

Photo/Drawing

Notes

BUG IDENTIFICATION

SEASON	
Spring	☐
Summer	☐
Fall	☐
Winter	☐

WEATHER	
Sunny	☐
Rainy	☐
Foggy	☐
Cloudy	☐

Bug Name:	
Scientific Name:	
Location:	
Colors:	
Number of Legs:	
Does It Have Wings?	
Behavior:	
Alone or in Group?	

Photo/Drawing

Notes

BUG IDENTIFICATION

SEASON	
Spring	☐
Summer	☐
Fall	☐
Winter	☐

WEATHER	
Sunny	☐
Rainy	☐
Foggy	☐
Cloudy	☐

Bug Name:	
Scientific Name:	
Location:	
Colors:	
Number of Legs:	
Does It Have Wings?	
Behavior:	
Alone or in Group?	

Photo/Drawing

Notes

BUG IDENTIFICATION

SEASON

- Spring ☐
- Summer ☐
- Fall ☐
- Winter ☐

WEATHER

- Sunny ☐
- Rainy ☐
- Foggy ☐
- Cloudy ☐

Bug Name:	
Scientific Name:	
Location:	
Colors:	
Number of Legs:	
Does It Have Wings?	
Behavior:	
Alone or in Group?	

Photo/Drawing

Notes

BUG IDENTIFICATION

SEASON	
Spring	☐
Summer	☐
Fall	☐
Winter	☐

WEATHER	
Sunny	☐
Rainy	☐
Foggy	☐
Cloudy	☐

Bug Name:	
Scientific Name:	
Location:	
Colors:	
Number of Legs:	
Does It Have Wings?	
Behavior:	
Alone or in Group?	

PHOTO/DRAWING

NOTES

Bug Identification

Season	
Spring	☐
Summer	☐
Fall	☐
Winter	☐

Weather	
Sunny	☐
Rainy	☐
Foggy	☐
Cloudy	☐

Bug Name:	
Scientific Name:	
Location:	
Colors:	
Number of Legs:	
Does It Have Wings?	
Behavior:	
Alone or in Group?	

PHOTO/DRAWING

NOTES

BUG IDENTIFICATION

SEASON	
Spring	☐
Summer	☐
Fall	☐
Winter	☐

WEATHER	
Sunny	☐
Rainy	☐
Foggy	☐
Cloudy	☐

Bug Name:	
Scientific Name:	
Location:	
Colors:	
Number of Legs:	
Does It Have Wings?	
Behavior:	
Alone or in Group?	

Photo/Drawing

Notes

BUG IDENTIFICATION

SEASON

- Spring ☐
- Summer ☐
- Fall ☐
- Winter ☐

WEATHER

- Sunny ☐
- Rainy ☐
- Foggy ☐
- Cloudy ☐

Bug Name:	
Scientific Name:	
Location:	
Colors:	
Number of Legs:	
Does It Have Wings?	
Behavior:	
Alone or in Group?	

PHOTO/DRAWING

NOTES

BUG IDENTIFICATION

Season	
Spring	☐
Summer	☐
Fall	☐
Winter	☐

Weather	
Sunny	☐
Rainy	☐
Foggy	☐
Cloudy	☐

Bug Name:	
Scientific Name:	
Location:	
Colors:	
Number of Legs:	
Does It Have Wings?	
Behavior:	
Alone or in Group?	

Photo/Drawing

Notes

Bug Identification

Season	
Spring	☐
Summer	☐
Fall	☐
Winter	☐

Weather	
Sunny	☐
Rainy	☐
Foggy	☐
Cloudy	☐

Bug Name:	
Scientific Name:	
Location:	
Colors:	
Number of Legs:	
Does It Have Wings?	
Behavior:	
Alone or in Group?	

PHOTO/DRAWING

NOTES

Bug Identification

Season	
Spring	☐
Summer	☐
Fall	☐
Winter	☐

Weather	
Sunny	☐
Rainy	☐
Foggy	☐
Cloudy	☐

Bug Name:	
Scientific Name:	
Location:	
Colors:	
Number of Legs:	
Does It Have Wings?	
Behavior:	
Alone or in Group?	

PHOTO/DRAWING

NOTES

BUG IDENTIFICATION

SEASON	
Spring	☐
Summer	☐
Fall	☐
Winter	☐

WEATHER	
Sunny	☐
Rainy	☐
Foggy	☐
Cloudy	☐

Bug Name:	
Scientific Name:	
Location:	
Colors:	
Number of Legs:	
Does It Have Wings?	
Behavior:	
Alone or in Group?	

PHOTO/DRAWING

NOTES

Bug Identification

Season	
Spring	☐
Summer	☐
Fall	☐
Winter	☐

Weather	
Sunny	☐
Rainy	☐
Foggy	☐
Cloudy	☐

Bug Name:	
Scientific Name:	
Location:	
Colors:	
Number of Legs:	
Does It Have Wings?	
Behavior:	
Alone or in Group?	

PHOTO/DRAWING

NOTES

BUG IDENTIFICATION

SEASON

- Spring ☐
- Summer ☐
- Fall ☐
- Winter ☐

WEATHER

- Sunny ☐
- Rainy ☐
- Foggy ☐
- Cloudy ☐

Bug Name:	
Scientific Name:	
Location:	
Colors:	
Number of Legs:	
Does It Have Wings?	
Behavior:	
Alone or in Group?	

Photo/Drawing

Notes

BUG IDENTIFICATION

SEASON	
Spring	☐
Summer	☐
Fall	☐
Winter	☐

WEATHER	
Sunny	☐
Rainy	☐
Foggy	☐
Cloudy	☐

Bug Name:	
Scientific Name:	
Location:	
Colors:	
Number of Legs:	
Does It Have Wings?	
Behavior:	
Alone or in Group?	

Photo/Drawing

Notes

BUG IDENTIFICATION

SEASON	
Spring	☐
Summer	☐
Fall	☐
Winter	☐

WEATHER	
Sunny	☐
Rainy	☐
Foggy	☐
Cloudy	☐

Bug Name:	
Scientific Name:	
Location:	
Colors:	
Number of Legs:	
Does It Have Wings?	
Behavior:	
Alone or in Group?	

PHOTO/DRAWING

NOTES

BUG IDENTIFICATION

SEASON		WEATHER	
Spring	☐	Sunny	☐
Summer	☐	Rainy	☐
Fall	☐	Foggy	☐
Winter	☐	Cloudy	☐

Bug Name:	
Scientific Name:	
Location:	
Colors:	
Number of Legs:	
Does It Have Wings?	
Behavior:	
Alone or in Group?	

PHOTO/DRAWING

NOTES

BUG IDENTIFICATION

SEASON	
Spring	☐
Summer	☐
Fall	☐
Winter	☐

WEATHER	
Sunny	☐
Rainy	☐
Foggy	☐
Cloudy	☐

Bug Name:	
Scientific Name:	
Location:	
Colors:	
Number of Legs:	
Does It Have Wings?	
Behavior:	
Alone or in Group?	

Photo/Drawing

Notes

Bug Identification

Season
- [ ] Spring
- [ ] Summer
- [ ] Fall
- [ ] Winter

Weather
- [ ] Sunny
- [ ] Rainy
- [ ] Foggy
- [ ] Cloudy

Bug Name:	
Scientific Name:	
Location:	
Colors:	
Number of Legs:	
Does It Have Wings?	
Behavior:	
Alone or in Group?	

PHOTO/DRAWING

NOTES

BUG IDENTIFICATION

SEASON	
Spring	☐
Summer	☐
Fall	☐
Winter	☐

WEATHER	
Sunny	☐
Rainy	☐
Foggy	☐
Cloudy	☐

Bug Name:	
Scientific Name:	
Location:	
Colors:	
Number of Legs:	
Does It Have Wings?	
Behavior:	
Alone or in Group?	

PHOTO/DRAWING

NOTES

Bug Identification

Season
- Spring ☐
- Summer ☐
- Fall ☐
- Winter ☐

Weather
- Sunny ☐
- Rainy ☐
- Foggy ☐
- Cloudy ☐

Bug Name:	
Scientific Name:	
Location:	
Colors:	
Number of Legs:	
Does It Have Wings?	
Behavior:	
Alone or in Group?	

PHOTO/DRAWING

NOTES

BUG IDENTIFICATION

SEASON	
Spring	☐
Summer	☐
Fall	☐
Winter	☐

WEATHER	
Sunny	☐
Rainy	☐
Foggy	☐
Cloudy	☐

Bug Name:	
Scientific Name:	
Location:	
Colors:	
Number of Legs:	
Does It Have Wings?	
Behavior:	
Alone or in Group?	

Photo/Drawing

Notes

BUG IDENTIFICATION

SEASON

- Spring ☐
- Summer ☐
- Fall ☐
- Winter ☐

WEATHER

- Sunny ☐
- Rainy ☐
- Foggy ☐
- Cloudy ☐

Bug Name:	
Scientific Name:	
Location:	
Colors:	
Number of Legs:	
Does It Have Wings?	
Behavior:	
Alone or in Group?	

PHOTO/DRAWING

NOTES

Bug Identification

Season	
Spring	☐
Summer	☐
Fall	☐
Winter	☐

Weather	
Sunny	☐
Rainy	☐
Foggy	☐
Cloudy	☐

Bug Name:	
Scientific Name:	
Location:	
Colors:	
Number of Legs:	
Does It Have Wings?	
Behavior:	
Alone or in Group?	

PHOTO/DRAWING

NOTES

Bug Identification

Season
- Spring ☐
- Summer ☐
- Fall ☐
- Winter ☐

Weather
- Sunny ☐
- Rainy ☐
- Foggy ☐
- Cloudy ☐

Bug Name:	
Scientific Name:	
Location:	
Colors:	
Number of Legs:	
Does It Have Wings?	
Behavior:	
Alone or in Group?	

PHOTO/DRAWING

NOTES

BUG IDENTIFICATION

SEASON

- Spring ☐
- Summer ☐
- Fall ☐
- Winter ☐

WEATHER

- Sunny ☐
- Rainy ☐
- Foggy ☐
- Cloudy ☐

Bug Name:	
Scientific Name:	
Location:	
Colors:	
Number of Legs:	
Does It Have Wings?	
Behavior:	
Alone or in Group?	

Photo/Drawing

Notes

BUG IDENTIFICATION

SEASON

- Spring ☐
- Summer ☐
- Fall ☐
- Winter ☐

WEATHER

- Sunny ☐
- Rainy ☐
- Foggy ☐
- Cloudy ☐

Bug Name:	
Scientific Name:	
Location:	
Colors:	
Number of Legs:	
Does It Have Wings?	
Behavior:	
Alone or in Group?	

PHOTO/DRAWING

NOTES

BUG IDENTIFICATION

SEASON

- Spring ☐
- Summer ☐
- Fall ☐
- Winter ☐

WEATHER

- Sunny ☐
- Rainy ☐
- Foggy ☐
- Cloudy ☐

Bug Name:	
Scientific Name:	
Location:	
Colors:	
Number of Legs:	
Does It Have Wings?	
Behavior:	
Alone or in Group?	

PHOTO/DRAWING

NOTES

BUG IDENTIFICATION

SEASON

- Spring ☐
- Summer ☐
- Fall ☐
- Winter ☐

WEATHER

- Sunny ☐
- Rainy ☐
- Foggy ☐
- Cloudy ☐

Bug Name:	
Scientific Name:	
Location:	
Colors:	
Number of Legs:	
Does It Have Wings?	
Behavior:	
Alone or in Group?	

Photo/Drawing

Notes

Bug Identification

Season
- Spring ☐
- Summer ☐
- Fall ☐
- Winter ☐

Weather
- Sunny ☐
- Rainy ☐
- Foggy ☐
- Cloudy ☐

Bug Name:	
Scientific Name:	
Location:	
Colors:	
Number of Legs:	
Does It Have Wings?	
Behavior:	
Alone or in Group?	

PHOTO/DRAWING

NOTES

Bug Identification

Season
- Spring ☐
- Summer ☐
- Fall ☐
- Winter ☐

Weather
- Sunny ☐
- Rainy ☐
- Foggy ☐
- Cloudy ☐

Bug Name:	
Scientific Name:	
Location:	
Colors:	
Number of Legs:	
Does It Have Wings?	
Behavior:	
Alone or in Group?	

Photo/Drawing

Notes

BUG IDENTIFICATION

SEASON

- Spring ☐
- Summer ☐
- Fall ☐
- Winter ☐

WEATHER

- Sunny ☐
- Rainy ☐
- Foggy ☐
- Cloudy ☐

Bug Name:	
Scientific Name:	
Location:	
Colors:	
Number of Legs:	
Does It Have Wings?	
Behavior:	
Alone or in Group?	

Photo/Drawing

Notes

BUG IDENTIFICATION

SEASON

- Spring ☐
- Summer ☐
- Fall ☐
- Winter ☐

WEATHER

- Sunny ☐
- Rainy ☐
- Foggy ☐
- Cloudy ☐

Bug Name:	
Scientific Name:	
Location:	
Colors:	
Number of Legs:	
Does It Have Wings?	
Behavior:	
Alone or in Group?	

PHOTO/DRAWING

NOTES

BUG IDENTIFICATION

SEASON	
Spring	☐
Summer	☐
Fall	☐
Winter	☐

WEATHER	
Sunny	☐
Rainy	☐
Foggy	☐
Cloudy	☐

Bug Name:	
Scientific Name:	
Location:	
Colors:	
Number of Legs:	
Does It Have Wings?	
Behavior:	
Alone or in Group?	

PHOTO/DRAWING

NOTES

BUG IDENTIFICATION

SEASON	
Spring	☐
Summer	☐
Fall	☐
Winter	☐

WEATHER	
Sunny	☐
Rainy	☐
Foggy	☐
Cloudy	☐

Bug Name:	
Scientific Name:	
Location:	
Colors:	
Number of Legs:	
Does It Have Wings?	
Behavior:	
Alone or in Group?	

PHOTO/DRAWING

NOTES

BUG IDENTIFICATION

SEASON	
Spring	☐
Summer	☐
Fall	☐
Winter	☐

WEATHER	
Sunny	☐
Rainy	☐
Foggy	☐
Cloudy	☐

Bug Name:	
Scientific Name:	
Location:	
Colors:	
Number of Legs:	
Does It Have Wings?	
Behavior:	
Alone or in Group?	

PHOTO/DRAWING

NOTES

BUG IDENTIFICATION

SEASON	
Spring	☐
Summer	☐
Fall	☐
Winter	☐

WEATHER	
Sunny	☐
Rainy	☐
Foggy	☐
Cloudy	☐

Bug Name:	
Scientific Name:	
Location:	
Colors:	
Number of Legs:	
Does It Have Wings?	
Behavior:	
Alone or in Group?	

PHOTO/DRAWING

NOTES

BUG IDENTIFICATION

SEASON

- Spring ☐
- Summer ☐
- Fall ☐
- Winter ☐

WEATHER

- Sunny ☐
- Rainy ☐
- Foggy ☐
- Cloudy ☐

Bug Name:	
Scientific Name:	
Location:	
Colors:	
Number of Legs:	
Does It Have Wings?	
Behavior:	
Alone or in Group?	

Photo/Drawing

Notes

BUG IDENTIFICATION

SEASON	
Spring	☐
Summer	☐
Fall	☐
Winter	☐

WEATHER	
Sunny	☐
Rainy	☐
Foggy	☐
Cloudy	☐

Bug Name:	
Scientific Name:	
Location:	
Colors:	
Number of Legs:	
Does It Have Wings?	
Behavior:	
Alone or in Group?	

Photo/Drawing

Notes

BUG IDENTIFICATION

SEASON	
Spring	☐
Summer	☐
Fall	☐
Winter	☐

WEATHER	
Sunny	☐
Rainy	☐
Foggy	☐
Cloudy	☐

Bug Name:	
Scientific Name:	
Location:	
Colors:	
Number of Legs:	
Does It Have Wings?	
Behavior:	
Alone or in Group?	

PHOTO/DRAWING

NOTES

BUG IDENTIFICATION

SEASON

- Spring ☐
- Summer ☐
- Fall ☐
- Winter ☐

WEATHER

- Sunny ☐
- Rainy ☐
- Foggy ☐
- Cloudy ☐

Bug Name:	
Scientific Name:	
Location:	
Colors:	
Number of Legs:	
Does It Have Wings?	
Behavior:	
Alone or in Group?	

Photo/Drawing

Notes

Bug Identification

Season	
Spring	☐
Summer	☐
Fall	☐
Winter	☐

Weather	
Sunny	☐
Rainy	☐
Foggy	☐
Cloudy	☐

Bug Name:	
Scientific Name:	
Location:	
Colors:	
Number of Legs:	
Does It Have Wings?	
Behavior:	
Alone or in Group?	

Photo/Drawing

Notes

BUG IDENTIFICATION

SEASON

- Spring ☐
- Summer ☐
- Fall ☐
- Winter ☐

WEATHER

- Sunny ☐
- Rainy ☐
- Foggy ☐
- Cloudy ☐

Bug Name:	
Scientific Name:	
Location:	
Colors:	
Number of Legs:	
Does It Have Wings?	
Behavior:	
Alone or in Group?	

Photo/Drawing

Notes

BUG IDENTIFICATION

SEASON

- Spring ☐
- Summer ☐
- Fall ☐
- Winter ☐

WEATHER

- Sunny ☐
- Rainy ☐
- Foggy ☐
- Cloudy ☐

Bug Name:	
Scientific Name:	
Location:	
Colors:	
Number of Legs:	
Does It Have Wings?	
Behavior:	
Alone or in Group?	

PHOTO/DRAWING

NOTES

BUG IDENTIFICATION

SEASON

- Spring ☐
- Summer ☐
- Fall ☐
- Winter ☐

WEATHER

- Sunny ☐
- Rainy ☐
- Foggy ☐
- Cloudy ☐

Bug Name:	
Scientific Name:	
Location:	
Colors:	
Number of Legs:	
Does It Have Wings?	
Behavior:	
Alone or in Group?	

Photo/Drawing

Notes

BUG IDENTIFICATION

SEASON	
Spring	☐
Summer	☐
Fall	☐
Winter	☐

WEATHER	
Sunny	☐
Rainy	☐
Foggy	☐
Cloudy	☐

Bug Name:	
Scientific Name:	
Location:	
Colors:	
Number of Legs:	
Does It Have Wings?	
Behavior:	
Alone or in Group?	

PHOTO/DRAWING

NOTES

BUG IDENTIFICATION

SEASON

- Spring ☐
- Summer ☐
- Fall ☐
- Winter ☐

WEATHER

- Sunny ☐
- Rainy ☐
- Foggy ☐
- Cloudy ☐

Bug Name:	
Scientific Name:	
Location:	
Colors:	
Number of Legs:	
Does It Have Wings?	
Behavior:	
Alone or in Group?	

PHOTO/DRAWING

NOTES

BUG IDENTIFICATION

SEASON	
Spring	☐
Summer	☐
Fall	☐
Winter	☐

WEATHER	
Sunny	☐
Rainy	☐
Foggy	☐
Cloudy	☐

Bug Name:	
Scientific Name:	
Location:	
Colors:	
Number of Legs:	
Does It Have Wings?	
Behavior:	
Alone or in Group?	

PHOTO/DRAWING

NOTES

BUG IDENTIFICATION

SEASON

- Spring ☐
- Summer ☐
- Fall ☐
- Winter ☐

WEATHER

- Sunny ☐
- Rainy ☐
- Foggy ☐
- Cloudy ☐

Bug Name:	
Scientific Name:	
Location:	
Colors:	
Number of Legs:	
Does It Have Wings?	
Behavior:	
Alone or in Group?	

Photo/Drawing

Notes

let's color

www.ingramcontent.com/pod-product-compliance
Lightning Source LLC
Chambersburg PA
CBHW060848220526
45466CB00003B/1286